EMPEROR PENGUINS

A MIGRATION STORY

amicus
LEARNING

by Lisa Amstutz ▪ illustrated by Howard Gray

AMICUS ILLUSTRATED is published by Amicus Learning, an imprint of Amcius
P.O. Box 227, Mankato, MN 56002
www.amicuspublishing.us

Editor: Alissa Thielges
Series Designer: Kim Pfeffer
Book Designer: Emily Dietz

Library of Congress Cataloging-in-Publication Data
Names: Amstutz, Lisa J., author. | Gray, Howard, illustrator.
Title: Emperor penguins : a migration story / by Lisa Amstutz; illustrated by Howard Gray.
Description: Mankato, MN : Amicus Illustrated, [2026] | Series: Incredible migrations | Includes bibliographical references. | Audience: Ages 5–10 | Audience: Grades 2–3 | Summary: "Follow the incredible journey of Emperor penguins' migration across Antarctica in this narrative nonfiction picture book that will delight animal and nature lovers and support life science education. Includes a migration map, tips to protect animals and habitats, a glossary, and further resources"—Provided by publisher.
Identifiers: LCCN 2024043399 (print) | LCCN 2024043400 (ebook) | ISBN 9781645492825 (library binding) | ISBN 9781681528069 (paperback) | ISBN 9781645493709 (ebook)
Subjects: LCSH: Emperor penguin—Juvenile literature. | Emperor penguin—Migration—Juvenile literature.
Classification: LCC QL696.S473 A467 2026 (print) | LCC QL696.S473 (ebook) | DDC 598.47156/8—dc23/eng/20250107
LC record available at https://lccn.loc.gov/2024043399
LC ebook record available at https://lccn.loc.gov/2024043400

About the Author

Lisa Amstutz is the author of more than 150 children's books. A former outdoor educator, she holds degrees in biology and environmental science. Lisa enjoys learning fun facts about science and sharing them with kids. She lives on a small farm with her family.

About the Illustrator

Howard Gray has illustrated a selection of fiction and non-fiction children's books. He has always considered himself an artist, but with a PhD in dolphin genetics, he has a background in zoology. He is now pursuing his dream career in children's illustration from the picturesque city of Durham, UK. Find out more at www.howardgrayillustrations.com.

Whoosh! Icy March winds blow. It is fall in Antarctica. An Emperor penguin snaps up a fish along the coast. It will be her last meal for months. A long journey lies ahead. She must reach her breeding grounds by winter. There, she will find a mate and lay an egg.

Penguins gather for the long march. They will migrate up to 100 miles (161 kilometers) inland. There, the thick ice won't melt until November. Penguin chicks will have time to grow.

The penguin walks and slides on the ice. The temperature dips to -40 degrees Fahrenheit (-40 degrees Celsius). The penguin's feathers and fat keep her warm.

At last, the penguin reaches the breeding grounds. *Squawk! Honk!* Noisy calls fill the air. She starts looking for a mate right away. She listens to the unique calls. When she finds a mate, they bow and call to each other. Then they huddle together.

In May, the penguin lays one egg. Her mate rolls it onto his feet. He tucks it under his brood pouch. This patch of bare skin keeps the egg warm. He will protect the egg as the chick grows inside.

Now the female heads back to sea. She must find food soon. Her chick will need to eat.

It takes weeks to reach open water. There, she dives for fish, squid, and krill. She watches for seals, sea lions, and orcas. These animals feed on penguins.

The male stays close to others to keep warm. They take turns moving to the middle of the group. More than two months pass.

One day, the egg moves. *Tap, tap! Crack!* A gray chick hatches out. The male sets it on his feet and covers it with his brood pouch. He feeds it crop milk. But it will need more food soon.

Just in time, the female returns. She listens for her mate's call. She finds him in the crowd. The penguin spits up food to feed her chick. Then she tucks the chick onto her feet. Now it is her turn to wait.

The male goes to find food. It has been four months since his last meal. He has lost nearly half of his body weight. He feeds for a few weeks. Then he returns with food for the chick.

Finding enough food can be hard. Fishing boats catch krill near the Antarctic coast. Penguins travel farther away to find food. While they're away, the chick could starve or be eaten by other animals.

Months pass. The sea ice begins to melt and break up. The ocean is nearer now. Mom and Dad keep busy finding food. The chick grows fast. By September, the chick can stay on its own. It huddles with others. The chick cannot swim until it molts. If the sea ice breaks up too soon, it may drown. Climate change is causing this to happen more often.

In December, the birds return to open sea. They spend the Antarctic summer months feeding. They store up as much fat as they can . . .

for another long, cold journey ahead.

EMPEROR PENGUIN MIGRATION

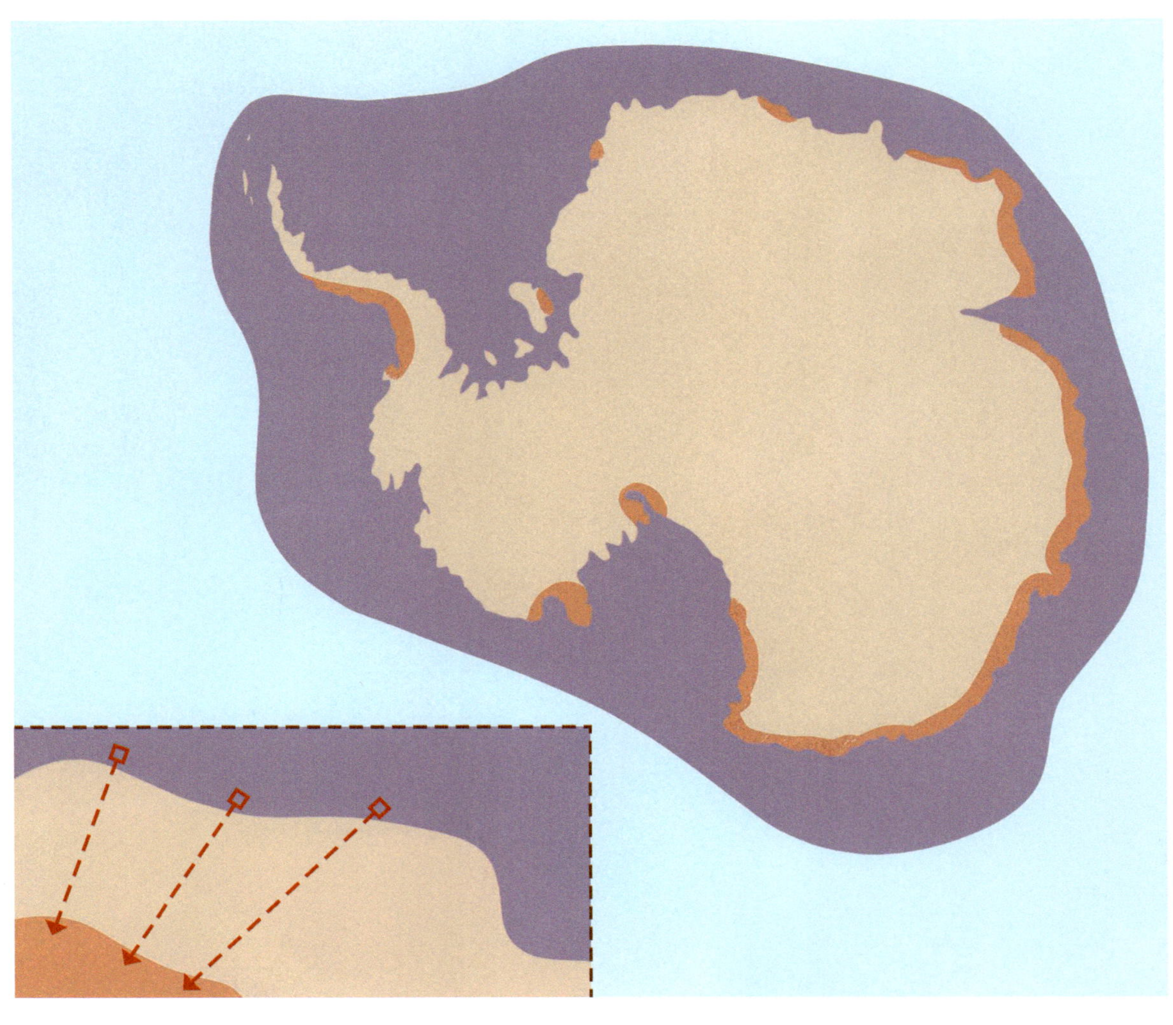

Climate change is caused by burning fossil fuels. Humans can help penguins by slowing climate change.

- Walk or bike instead of driving when you can.

- Turn off lights and devices when you aren't using them.

- Recycle or reuse items instead of buying new ones.

- Protect penguins' food supply from overfishing. Buy seafood that is caught without harming their habitat. The Monterey Bay Aquarium's Seafood Watch provides a helpful guide: https://www.seafoodwatch.org/recommendations/download-consumer-guides

Glossary

brood pouch A patch of bare skin that keeps the egg warm.

climate change A long-term shift in global or regional weather patterns.

crop milk A food made in a penguin's crop, a sac in the throat where it stores food.

krill Small shrimp-like animals that live in the ocean.

mate The female or male partner of a pair of animals.

migrate To move to a different place at a certain time of year, usually because of changing weather and seasons.

molt In birds, to shed a layer of feathers so that new ones can grow.

Read More

Banks, Rosie. ***Why Do Animals Migrate?*** New York: Gareth Stevens Publishing, 2024.

Davies, Nicola. ***Emperor of the Ice: How a Changing Climate Affects a Penguin Color*** Somerville, Mass.: Candlewick Press, 2023.

Saszaklis, John. ***My Life as an Emperor Penguin.*** North Mankato, Minn.: Picture Window Books, 2022.

Websites

Emperor Penguin
https://kids.nationalgeographic.com/animals/birds/emperor-penguin/

How Does Huddling Help Emperor Penguins Stay Warm?
https://thekidshouldseethis.com/post/penguin-huddling-social-thermoregulation

Top 10 Facts about Emperor Penguins
https://wwf.ca/stories/10-facts-about-emperor-penguins/

Every effort has been made to ensure that these websites are appropriate for children. However, because of the nature of the Internet, it is impossible to guarantee that these sites will remain active indefinitely or that their contents will not be altered.